Math Formulas and Equations for Students

350 Essential Mathematical Formulas and Equations

by

Peter I. Kattan

Petra Books
www.PetraBooks.com

Peter I. Kattan, PhD

Correspondence about this book may be sent to the author at one of the following two email addresses:

pkattan@petrabooks.com

info@petrabooks.com

The author acknowledges the work of Anubislivess which appears in the image of Pi on the front and back covers of this book.

Math Formulas and Equations for Students:
350 Essential Mathematical Formulas and Equations.
Written by Peter I. Kattan.
ISBN: 979-8-8690-8966-3

To my parents, brothers, and sisters

Math Formulas and Equations for Students
350 Essential Mathematical Formulas and Equations

Preface

This is a little book for students who wish to have the essential formulas and equations of mathematics in a single easily accessible source. In about 50 pages, the 350 most essential mathematical formulas are listed. Unlike other large books on this topic, there is no need to go through hundreds of pages and thousands of formulas for the student to get the basic equations. The author has searched several books on mathematical formulas and tables and selected only those equations which are essential to the student.

The mathematical formulas and equations listed in this book are useful for students and researchers in various fields including mathematics, physics, engineering, etc. Only the most elementary and basic topics are covered including formulas for various geometric shapes, several types of functions (trigonometric, hyperbolic, exponential, logarithmic, etc), the quadratic equation, analytic geometry, derivatives and integrals, arithmetic series, geometric series, and Taylor series.

A comprehensive reference list is included at the end of the book in addition to numerous web links to more formulas, equations, and tables. The author has decided against including numerical tables for integration, logarithms, etc because the use of such tables has been superseded by the availability of scientific calculators and computer algebra programs. Thus, a decision has been reached to keep the book in a compact format that includes only the most essential and basic formulas. It is hoped that the author has succeeded in this regard.

Extreme care has been taken to ensure the accuracy of the formulas that are listed in this book. It is hoped that this little book will prove to be a valuable source of information for students in mathematics, science, and engineering. Finally, the author wishes to acknowledge the help and support of his family members without which he would not have been able to produce this book in its present form.

Peter I. Kattan
January 2024

Contents

1. Special Constants

$\pi = 3.14159$ Ratio of perimeter of a circle to its diameter

$e = 2.71828$ Natural base of logarithms

$\gamma = 0.57721$ Euler's constant

$\sqrt{2} = 1.41421$ Square root of 2

$\sqrt{3} = 1.73205$ Square root of 3

$\sqrt{5} = 2.23607$ Square root of 5

2. Special Products and Factors

$$(x + y)^2 = x^2 + 2xy + y^2$$

$$(x - y)^2 = x^2 - 2xy + y^2$$

$$(x + y)^3 = x^3 + 3x^2y + 3xy^2 + y^3$$

$$(x - y)^3 = x^3 - 3x^2y + 3xy^2 - y^3$$

$$(x + y)^4 = x^4 + 4x^3y + 6x^2y^2 + 4xy^3 + y^4$$

$$(x - y)^4 = x^4 - 4x^3y + 6x^2y^2 - 4xy^3 + y^4$$

$$x^2 - y^2 = (x - y)(x + y)$$

$$x^2 + y^2 = (x - iy)(x + iy) \quad \text{Factors are complex numbers}$$
$$\text{(cannot be factored with real factors)}$$

$$x^3 - y^3 = (x - y)(x^2 + xy + y^2)$$

$$x^3 + y^3 = (x + y)(x^2 - xy + y^2)$$

$$x^4 - y^4 = (x^2 - y^2)(x^2 + y^2)$$
$$= (x - y)(x + y)(x^2 + y^2)$$
$$= (x - y)(x + y)(x - iy)(x + iy)$$
$$\text{The last line above includes complex factors.}$$

3. Binomial Formula and Binomial Coefficients

$$(x + y)^n = x^n + \binom{n}{1}x^{n-1}y + \binom{n}{2}x^{n-2}y^2 + \binom{n}{3}x^{n-3}y^3 + \ldots + \binom{n}{n}y^n$$

where the binomial coefficients are given by

$$\binom{n}{k} = \frac{n!}{k!(n-k)!} \quad \text{and} \quad n! = 1 \cdot 2 \cdot 3 \cdot 4 \cdots (n-1) \cdot n$$

for any integer n, and $0! = 1$.

Note that $\binom{n}{k} = \binom{n}{n-k}$

4. Geometric Formulas

Areas of Common Two-Dimensional Shapes

Shape	Area
Square of side a	$Area = a^2$
Rectangle of length a and width b	$Area = ab$
Triangle of altitude h and base b	$Area = \dfrac{1}{2}bh$
Triangle of sides a, b, and c	$Area = \sqrt{s(s-a)(s-b)(s-c)}$ $where \ s = \dfrac{1}{2}(a+b+c)$
Trapezoid of altitude h and parallel sides a and b	$Area = \dfrac{1}{2}h(a+b)$
Circle of radius r	$Area = \pi r^2$
Circle of diameter d	$Area = \dfrac{1}{4}\pi d^2$
Sector of circle with radius r and angle θ	$Area = \dfrac{1}{2}r^2\theta$, $\theta \ in \ radians$
Ellipse of semi-major axis a and semi-minor-axis b	$Area = \pi ab$

Perimeters of Common Two-Dimensional Shapes

Shape	Perimeter
Square of side a	$Perimeter = 4a$
Rectangle of length a and width b	$Perimeter = 2(a+b)$
Triangle of sides a, b, and c	$Perimeter = a+b+c$
Circle of radius r	$Perimeter = 2\pi r$

Circle of diameter d	$Perimeter = \pi d$
Sector of circle with radius r and angle θ	$Arc\,Length\ s = r\theta$, θ in radians

Volumes of Common Three-Dimensional Solids

Solid	Volume
Cube of side a	$Volume = a^3$
Rectangular parallelepiped of length a, width b, and height c	$Volume = abc$
Sphere of radius r	$Volume = \dfrac{4}{3}\pi r^3$
Right circular cylinder of radius r and height h	$Volume = \pi r^2 h$
Right circular cone of radius r and height h	$Volume = \dfrac{1}{3}\pi r^2 h$
Pyramid of base area A and height h	$Volume = \dfrac{1}{3}Ah$
Ellipsoid of semi-axes a, b, and c	$Volume = \dfrac{4}{3}\pi abc$

Surface Areas of Common Three-Dimensional Solids

Solid	Surface Area
Cube of side a	$Surface\,Area = 6a^2$
Rectangular parallelepiped of length a, width b, and height c	$Surface\,Area = 2(ab + ac + bc)$
Sphere of radius r	$Surface\,Area = 4\pi r^2$
Right circular cylinder of radius r and height h	$Lateral\,Surface\,Area = 2\pi r h$

Right circular cone of radius r and height h	$Lateral\ Surface\ Area = \pi r \sqrt{r^2 + h^2}$

5. Trigonometric Functions

Consider a right triangle with angle θ and sides x, y, and r.

$$\sin\theta = \frac{y}{r} = \frac{opposite}{hypotenuse}$$

$$\cos\theta = \frac{x}{r} = \frac{adjacent}{hypotenuse}$$

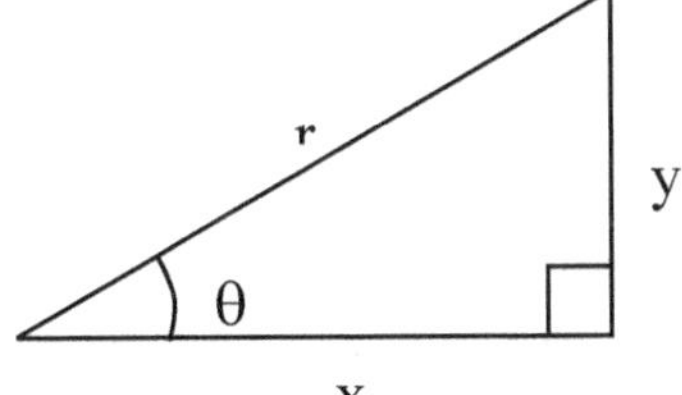

$$\tan\theta = \frac{y}{x} = \frac{opposite}{adjacent}$$

For angles, we have the following relations:

$$2\pi\ radians = 360^o$$

$$\pi\ radians = 180^o$$

$$1\ radians = \frac{180^o}{\pi}$$

$$1^o = \frac{\pi}{180^o}\ radians$$

We have also the following trigonometric identities:

$$\tan\theta = \frac{\sin\theta}{\cos\theta}$$

$$\cot\theta = \frac{\cos\theta}{\sin\theta} = \frac{1}{\tan\theta}$$

$$\sec\theta = \frac{1}{\cos\theta}$$

$$\csc\theta = \frac{1}{\sin\theta}$$

$$\sin^2\theta + \cos^2\theta = 1$$

$$\sec^2\theta = 1 + \tan^2\theta$$

$$\csc^2\theta = 1 + \cot^2\theta$$

We have the following results for some common angles:

θ (degrees)	θ (radians)	$\sin\theta$	$\cos\theta$
0°	0	0	1
30°	$\dfrac{\pi}{6}$	$\dfrac{1}{2}$	$\dfrac{\sqrt{3}}{2}$
45°	$\dfrac{\pi}{4}$	$\dfrac{\sqrt{2}}{2}$	$\dfrac{\sqrt{2}}{2}$
60°	$\dfrac{\pi}{3}$	$\dfrac{\sqrt{3}}{2}$	$\dfrac{1}{2}$
90°	$\dfrac{\pi}{2}$	1	0

Formulas for the negative of an angle

$$\sin(-\theta) = -\sin\theta$$

$$\cos(-\theta) = \cos\theta$$

$$\tan(-\theta) = -\tan\theta$$

Co-function formulas

$$\sin\left(\frac{\pi}{2} - \theta\right) = \cos\theta$$

$$\cos\left(\frac{\pi}{2} - \theta\right) = \sin\theta$$

Formulas for sums and differences of two angles

$$\sin(\alpha + \beta) = \sin\alpha\cos\beta + \cos\alpha\sin\beta$$

$$\sin(\alpha - \beta) = \sin\alpha\cos\beta - \cos\alpha\sin\beta$$

$$\cos(\alpha + \beta) = \cos\alpha\cos\beta - \sin\alpha\sin\beta$$

$$\cos(\alpha - \beta) = \cos\alpha\cos\beta + \sin\alpha\sin\beta$$

$$\tan(\alpha + \beta) = \frac{\tan\alpha + \tan\beta}{1 - \tan\alpha\tan\beta}$$

$$\tan(\alpha - \beta) = \frac{\tan\alpha - \tan\beta}{1 + \tan\alpha\tan\beta}$$

Formulas for double angles

$$\sin(2\theta) = 2\sin\theta\cos\theta$$

$$\cos 2\theta = \cos^2\theta - \sin^2\theta$$
$$= 1 - 2\sin^2\theta$$
$$= 2\cos^2\theta - 1$$

$$\tan 2\theta = \frac{2\tan\theta}{1 - \tan^2\theta}$$

Formulas for half angles

$$\sin\left(\frac{\theta}{2}\right) = \pm\sqrt{\frac{1 - \cos\theta}{2}}$$

$$\cos\left(\frac{\theta}{2}\right) = \pm\sqrt{\frac{1 + \cos\theta}{2}}$$

$$\tan\left(\frac{\theta}{2}\right) = \pm\sqrt{\frac{1 - \cos\theta}{1 + \cos\theta}}$$
$$= \frac{\sin\theta}{1 + \cos\theta}$$
$$= \frac{1 - \cos\theta}{\sin\theta}$$
$$= \csc\theta - \cot\theta$$

Square of the sine and cosine functions

$$\sin^2 \theta = \frac{1 - \cos(2\theta)}{2}$$

$$\cos^2 \theta = \frac{1 + \cos(2\theta)}{2}$$

Sums, differences, and products of the sine and cosine functions

$$\sin \alpha + \sin \beta = 2\sin\left(\frac{\alpha + \beta}{2}\right)\cos\left(\frac{\alpha - \beta}{2}\right)$$

$$\sin \alpha - \sin \beta = 2\cos\left(\frac{\alpha + \beta}{2}\right)\sin\left(\frac{\alpha - \beta}{2}\right)$$

$$\cos \alpha + \cos \beta = 2\cos\left(\frac{\alpha + \beta}{2}\right)\cos\left(\frac{\alpha - \beta}{2}\right)$$

$$\cos \alpha - \cos \beta = -2\sin\left(\frac{\alpha + \beta}{2}\right)\sin\left(\frac{\alpha - \beta}{2}\right)$$

$$\sin \alpha \sin \beta = \frac{\cos(\alpha - \beta) - \cos(\alpha + \beta)}{2}$$

$$\cos \alpha \cos \beta = \frac{\cos(\alpha - \beta) + \cos(\alpha + \beta)}{2}$$

$$\sin \alpha \cos \beta = \frac{\sin(\alpha - \beta) + \sin(\alpha + \beta)}{2}$$

Law of Sines

$$\frac{a}{\sin A} = \frac{b}{\sin B} = \frac{c}{\sin C}$$

Law of Cosines

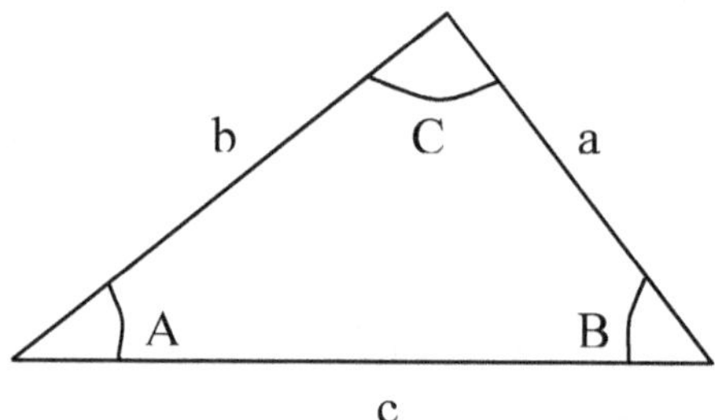

$$c = \sqrt{a^2 + b^2 - 2ab\cos C}$$

6. Complex Numbers

$$i = \sqrt{-1} \quad , \quad i^2 = -1$$

$a + bi$ is the general form of a complex number where a and b are real numbers.

$a + bi = c + di$ if and only if $a = c$ and $b = d$.

Addition and subtraction of complex numbers

$$(a + bi) + (c + di) = (a + c) + (b + d)i$$

$$(a + bi) - (c + di) = (a - c) + (b - d)i$$

Multiplication and division of complex numbers

$$(a + bi)(c + di) = (ac - bd) + (ad + bc)i$$

$$\frac{a+bi}{c+di} = \left(\frac{a+bi}{c+di}\right)\left(\frac{c-di}{c-di}\right) = \frac{(ac+bd)+(bc-ad)i}{c^2+d^2}$$

$$= \frac{ac+bd}{c^2+d^2} + \frac{bc-ad}{c^2+d^2}i$$

Polar form of a complex number

$$a+bi = r(\cos\theta + i\sin\theta)$$

where

$$r = \sqrt{a^2+b^2}$$

$$\theta = \tan^{-1}\left(\frac{b}{a}\right)$$

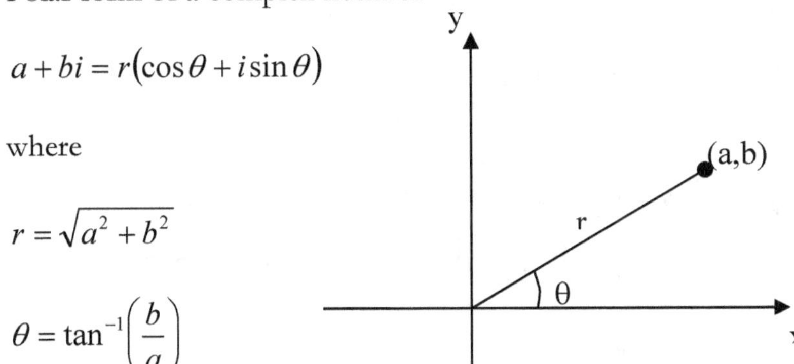

Multiplication and division of complex numbers in polar form

$$z_1 = r_1(\cos\theta_1 + i\sin\theta_1)$$

$$z_2 = r_2(\cos\theta_2 + i\sin\theta_2)$$

$$z_1 z_2 = r_1 r_2[\cos(\theta_1 + \theta_2) + i\sin(\theta_1 + \theta_2)]$$

$$\frac{z_1}{z_2} = \frac{r_1}{r_2}[\cos(\theta_1 - \theta_2) + i\sin(\theta_1 - \theta_2)]$$

Powers of complex numbers

$$z = r(\cos\theta + i\sin\theta)$$

$$z^n = r^n\left[\cos(n\theta) + i\sin(n\theta)\right]$$

Roots of complex numbers

$$z = r\left(\cos\theta + i\sin\theta\right)$$

$$\sqrt[n]{z} = z^{\frac{1}{n}} = r^{\frac{1}{n}}\left[\cos\left(\frac{\theta + 2k\pi}{n}\right) + i\sin\left(\frac{\theta + 2k\pi}{n}\right)\right]$$

where k is any integer.

7. Exponential and Logarithmic Functions

Rules for exponents

$$x^n = x \cdot x \cdot x \cdot x \ldots\ldots\ldots x \quad (n\ times) \quad , \quad x,n \ \text{are real numbers.}$$

$$x^m \cdot x^n = x^{m+n} \quad , \quad x,m,n \ \text{are real numbers.}$$

$$\frac{x^m}{x^n} = x^{m-n}$$

$$\left(x^m\right)^n = x^{mn}$$

$$x^0 = 1 \quad , \quad x \neq 0$$

$$x^{-n} = \frac{1}{x^n} \quad , \quad x \neq 0$$

$$(xy)^n = x^n y^n$$

$$\sqrt[n]{x} = x^{\frac{1}{n}}$$

$$\sqrt[n]{x^m} = x^{\frac{m}{n}}$$

$$\sqrt[n]{\frac{x}{y}} = \frac{\sqrt[n]{x}}{\sqrt[n]{y}}$$

Definition of logarithms

$$y = \log_a x \quad \text{if and only if} \quad x = a^y \quad , \quad a \neq 1$$

Properties of logarithms

$$\log_a a = 1$$

$$\log_a 1 = 0$$

$$\log_a(xy) = \log_a x + \log_a y$$

$$\log_a\left(\frac{x}{y}\right) = \log_a x - \log_a y$$

$$\log_a(x^n) = n\log_a x$$

Change of base for logarithms

$$\log_a x = \frac{\log_b x}{\log_b a} \quad , \quad a \neq 1 \ , \ b \neq 1$$

$$\log_a x = \frac{1}{\log_x a} \quad , \quad a \neq 1 \ , \ x \neq 1$$

Special logarithms

$$\ln x = \log_e x = \frac{\log_{10} x}{\log_{10} e} = \frac{\log_{10} x}{0.4343}$$

$$\log x = \log_{10} x = \frac{\ln x}{\ln 10} = \frac{\ln x}{2.3026}$$

Other formulas

$$e^{i\theta} = \cos\theta + i\sin\theta$$

$$e^{-i\theta} = \cos\theta - i\sin\theta$$

Relation between exponential and trigonometric functions

$$\sin\theta = \frac{e^{i\theta} - e^{-i\theta}}{2}$$

$$\cos\theta = \frac{e^{i\theta} + e^{-i\theta}}{2}$$

Periodicity of the exponential function

$$e^{i(\theta + 2k\pi)} = e^{i\theta} \quad , \quad k \text{ is any integer}$$

More on the polar form of complex numbers

$$z = x + iy = r(\cos\theta + i\sin\theta) = re^{i\theta}$$

Multiplication and division of complex numbers in polar form

$$z_1 = r_1 e^{i\theta_1}$$

$$z_2 = r_2 e^{i\theta_2}$$

$$z_1 z_2 = r_1 r_2 e^{i(\theta_1 + \theta_2)}$$

$$\frac{z_1}{z_2} = \frac{r_1}{r_2} e^{i(\theta_1 - \theta_2)}$$

Powers and roots of complex numbers in polar form

$$z = r e^{i\theta}$$

$$z^n = r^n e^{in\theta} \quad , \quad n \text{ is a real number}$$

$$z^{\frac{1}{n}} = r^{\frac{1}{n}} e^{i\left(\frac{\theta + 2k\pi}{n}\right)} \quad , \quad n, k \text{ are integers}$$

Logarithms of complex numbers in polar form

$$\ln z = \ln\left(r e^{i\theta}\right) = \ln r + i\theta + 2k\pi i, \quad k \text{ is an integer}$$

8. Hyperbolic Functions

$$\sinh x = \frac{e^x - e^{-x}}{2}$$

$$\cosh x = \frac{e^x + e^{-x}}{2}$$

$$\tanh x = \frac{e^x - e^{-x}}{e^x + e^{-x}} = \frac{\sinh x}{\cosh x}$$

$$\coth x = \frac{e^x + e^{-x}}{e^x - e^{-x}} = \frac{\cosh x}{\sinh x} = \frac{1}{\tanh x}$$

$$\sec h x = \frac{2}{e^x + e^{-x}} = \frac{1}{\cosh x}$$

$$\csc h x = \frac{2}{e^x - e^{-x}} = \frac{1}{\sinh x}$$

$$\cosh^2 x - \sinh^2 x = 1$$

$$\sec h^2 x + \tanh^2 x = 1$$

$$\coth^2 x - \csc h^2 x = 1$$

Relation between hyperbolic and trigonometric functions

$$\sin(ix) = i \sinh x$$

$$\cos(ix) = \cosh x$$

$$\tan(ix) = i \tanh x$$

$$\sinh(ix) = i \sin x$$

$$\cosh(ix) = \cos x$$

$$\tanh(ix) = i \tan x$$

9. The Quadratic Equation

The solution of the quadratic equation $ax^2 + bx + c = 0$ is given by the quadratic formula as follows (a,b,c are real)

$$x = \frac{-b \pm \sqrt{b^2 - 4ac}}{2a}$$

$D = b^2 - 4ac$ is a called the *discriminant*. We have the following three cases:

1. $D > 0$ implies there are two real but unequal solutions.
2. $D = 0$ implies there are two real and equal solutions.
3. $D < 0$ implies there are two complex conjugate solutions.

10. Plane Analytic Geometry

The distance d between two points with coordinates (x_1, y_1) and (x_2, y_2) is given by

$$d = \sqrt{(x_2 - x_1)^2 + (y_2 - y_1)^2}$$

The slope of the line joining two points with coordinates (x_1, y_1) and (x_2, y_2) is given by

$$m = \frac{y_2 - y_1}{x_2 - x_1} = \tan\theta$$

where θ is the angle of inclination of the line.

The equation of the line joining two points with coordinates (x_1, y_1) and (x_2, y_2) is given by

$$\frac{y - y_1}{x - x_1} = \frac{y_2 - y_1}{x_2 - x_1}$$

The equation of the line with slope m and passing through a point with coordinates (x_1, y_1) is given by

$$y - y_1 = m(x - x_1)$$

The equation of the line with slope m and y-intercept b is given by

$$y = mx + b$$

The equation of the line with x-intercept a and y-intercept b is given by

$$\frac{x}{a} + \frac{y}{b} = 1 \quad , \quad a \neq 0 \; , \; b \neq 0$$

The general equation of a line is given by

$$Ax + By + C = 0$$

where A, B, and C are constants.

The distance d from a point with coordinates (x_1, y_1) to the line given by the equation $Ax + By + C = 0$ is given by

$$d = \frac{Ax_1 + By_1 + C}{\pm \sqrt{A^2 + B^2}}$$

The angle θ between two lines with slopes m_1 and m_2 is given by

$$\theta = \tan^{-1}\left(\frac{m_2 - m_1}{1 + m_1 m_2} \right)$$

The area A of a triangle with vertices at the three points with coordinates (x_1, y_1), (x_2, y_2), and (x_3, y_3) is given by

$$A = \pm \frac{1}{2} \begin{vmatrix} x_1 & y_1 & 1 \\ x_2 & y_2 & 1 \\ x_3 & y_3 & 1 \end{vmatrix}$$

where the plus or minus sign is chosen to give a positive area.

Polar coordinates: The following are the relations between polar coordinates (r, θ) and rectangular (Cartesian) coordinates (x, y)

$$x = r \cos \theta$$

$$y = r \sin \theta$$

or alternatively

$$r = \sqrt{x^2 + y^2}$$

$$\theta = \tan^{-1}\left(\frac{y}{x}\right)$$

The equation of a circle with radius R and center at the point with coordinates (x_0, y_0) is given by

$$(x - x_0)^2 + (y - y_0)^2 = R^2$$

11. Solid Analytic Geometry

The distance d between two points with coordinates (x_1, y_1, z_1) and (x_2, y_2, z_2) is given by

$$d = \sqrt{(x_2 - x_1)^2 + (y_2 - y_1)^2 + (z_2 - z_1)^2}$$

The direction cosines l, m, and n of a line passing through two points with coordinates (x_1, y_1, z_1) and (x_2, y_2, z_2) are given by

$$l = \cos\alpha = \frac{x_2 - x_1}{d}$$

$$m = \cos\beta = \frac{y_2 - y_1}{d}$$

$$n = \cos\gamma = \frac{z_2 - z_1}{d}$$

where d is the distance between the two points and α, β, and γ are the angles of inclination of the line with the positive x, y, and z axes, respectively. We also have the following relation between the direction cosines:

$$\cos^2\alpha + \cos^2\beta + \cos^2\gamma = 1$$

or equivalently

$$l^2 + m^2 + n^2 = 1$$

The equations of a line passing through two points with coordinates (x_1, y_1, z_1) and (x_2, y_2, z_2) are given as follows in standard form

$$\frac{x - x_1}{x_2 - x_1} = \frac{y - y_1}{y_2 - y_1} = \frac{z - z_1}{z_2 - z_1}$$

or alternatively

$$\frac{x - x_1}{l} = \frac{y - y_1}{m} = \frac{z - z_1}{n}$$

The equations of a line passing through two points with coordinates (x_1, y_1, z_1) and (x_2, y_2, z_2) are given as follows in parametric form

$$x = x_1 + lt$$

$$y = y_1 + mt$$

$$z = z_1 + nt$$

where t is the parameter, and l, m, and n are the direction cosines.

The general equation of a plane is given by

$$Ax + By + Cz + D = 0$$

where A, B, C, and D are constants.

The equation of a plane passing through three points with coordinates (x_1, y_1, z_1), (x_2, y_2, z_2), and (x_3, y_3, z_3) is given by

$$\begin{vmatrix} x - x_1 & y - y_1 & z - z_1 \\ x_2 - x_1 & y_2 - y_1 & z_2 - z_1 \\ x_3 - x_1 & y_3 - y_1 & z_3 - z_1 \end{vmatrix} = 0$$

The equation of a line with x-intercept a, y-intercept b, and z-intercept c is given by

$$\frac{x}{a} + \frac{y}{b} + \frac{z}{c} = 1 \; , \quad a \neq 0 \; , \quad b \neq 0 \; , \quad c \neq 0$$

The equations of a line passing through a point with coordinates (x_0, y_0, z_0) and perpendicular to the plane $Ax + By + Cz + D = 0$ is given by in standard form

$$\frac{x - x_0}{A} = \frac{y - y_0}{B} = \frac{z - z_0}{C}$$

The equations of a line passing through a point with coordinates (x_0, y_0, z_0) and perpendicular to the plane $Ax + By + Cz + D = 0$ is given by in parametric form

$$x = x_0 + At$$
$$y = y_0 + Bt$$
$$z = z_0 + Ct$$

The distance d from a point with coordinates (x_0, y_0, z_0) to the plane $Ax + By + Cz + D = 0$ i z ven by

$$d = \frac{Ax_0 + By_0 + Cz_0 + D}{\pm\sqrt{A^2 + B^2 + C^2}}$$

Cylindrical coordinates (r, θ, z)

$$x = r\cos\theta$$
$$y = r\sin\theta$$
$$z = z$$

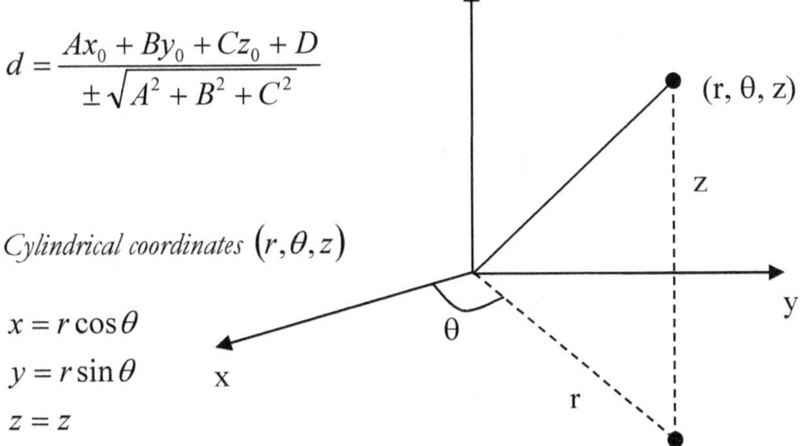

where (x, y, z) are the rectangular (Cartesian) coordinates. Alternatively, we have

$$r = \sqrt{x^2 + y^2}$$

$$\theta = \tan^{-1}\left(\frac{y}{x}\right)$$

$$z = z$$

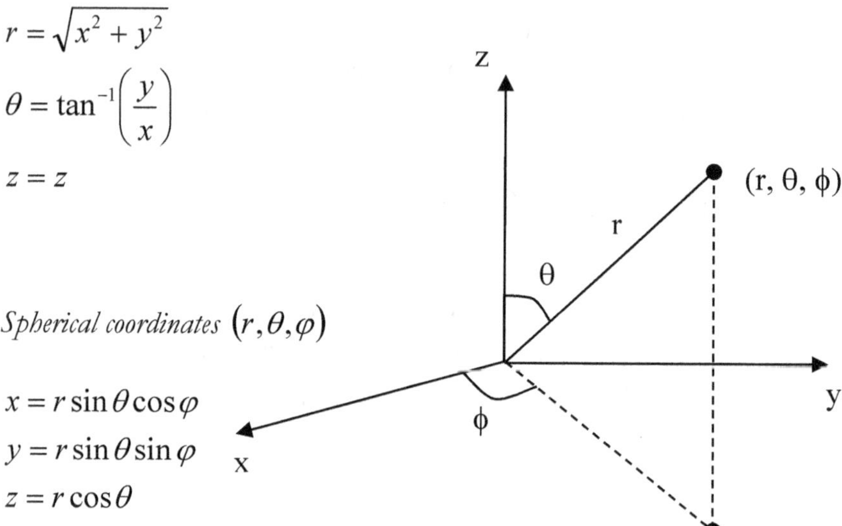

Spherical coordinates (r,θ,φ)

$$x = r\sin\theta\cos\varphi$$

$$y = r\sin\theta\sin\varphi$$

$$z = r\cos\theta$$

where (x,y,z) are the rectangular (Cartesian) coordinates. Alternatively, we have

$$r = \sqrt{x^2 + y^2 + z^2}$$

$$\varphi = \tan^{-1}\left(\frac{y}{x}\right)$$

$$\theta = \cos^{-1}\left(\frac{z}{\sqrt{x^2 + y^2 + z^2}}\right)$$

The equation of a sphere with radius R and center at a point with coordinates (x_0, y_0, z_0) is given by

$$(x - x_0)^2 + (y - y_0)^2 + (z - z_0)^2 = R^2$$

12. Derivatives

Definition of the derivative of a function.

Let $y = f(x)$ be a given function. Then its derivative is given by

$$f'(x) = \frac{dy}{dx} = \lim_{h \to 0} \frac{f(x+h) - f(x)}{h}$$

If we let $h = \Delta x$, then the above definition can be written equivalently as follows

$$f'(x) = \frac{dy}{dx} = \lim_{\Delta x \to 0} \frac{f(x + \Delta x) - f(x)}{\Delta x}$$

Rules of differentiation

In the following rules of differentiation, we assume $c = \text{constant}$, $n = \text{constant}$, and $u = f(x)$, $v = g(x)$

$$\frac{d}{dx}(c) = 0$$

$$\frac{d}{dx}(x) = 1$$

$$\frac{d}{dx}(cx) = c$$

$$\frac{d}{dx}\left(x^{n}\right) = nx^{n-1}$$

$$\frac{d}{dx}\left(cx^{n}\right) = ncx^{n-1}$$

$$\frac{d}{dx}\left(u+v\right) = \frac{du}{dx} + \frac{dv}{dx}$$

or equivalently $\qquad \left(f(x)+g(x)\right)' = f'(x) + g'(x)$

$$\frac{d}{dx}\left(u-v\right) = \frac{du}{dx} - \frac{dv}{dx}$$

or equivalently $\qquad \left(f(x)-g(x)\right)' = f'(x) - g'(x)$

$$\frac{d}{dx}\left(cu\right) = c\frac{du}{dx}$$

or equivalently $\qquad \left(cf(x)\right)' = cf'(x)$

$$\frac{d}{dx}\left(uv\right) = u\frac{dv}{dx} + v\frac{du}{dx}$$

or equivalently $\qquad \left(f(x)g(x)\right)' = f(x)g'(x) + g(x)f'(x)$

$$\frac{d}{dx}\left(\frac{u}{v}\right) = \frac{v\dfrac{du}{dx} - u\dfrac{dv}{dx}}{v^{2}} \quad , \quad v \neq 0$$

or equivalently

$$\left(\frac{f(x)}{g(x)}\right)' = \frac{g(x)f'(x) - f(x)g'(x)}{(g(x))^2} \quad,$$

$$g(x) \neq 0$$

$$\frac{d}{dx}\left(u^n\right) = nu^{n-1}\frac{du}{dx}$$

or equivalently

$$\left((f(x))^n\right)' = n(f(x))^{n-1}f'(x)$$

Chain rule

$$\frac{dy}{dx} = \frac{dy}{du}\frac{du}{dx}$$

or equivalently

$$(f(g(x)))' = f'(g(x))g'(x)$$

$$\frac{du}{dx} = \frac{1}{\left(\dfrac{dx}{du}\right)}$$

$$\frac{dy}{dx} = \frac{\left(\dfrac{dy}{du}\right)}{\left(\dfrac{dx}{du}\right)}$$

Derivatives of trigonometric functions

$$\frac{d}{dx}(\sin x) = \cos x$$

$$\frac{d}{dx}(\cos x) = -\sin x$$

$$\frac{d}{dx}(\tan x) = \sec^2 x$$

$$\frac{d}{dx}(\cot x) = -\csc^2 x$$

$$\frac{d}{dx}(\sec x) = \sec x \tan x$$

$$\frac{d}{dx}(\csc x) = -\csc x \cot x$$

$$\frac{d}{dx}(\sin u) = \cos u \frac{du}{dx}$$

or equivalently $\qquad (\sin(f(x)))' = \cos(f(x))f'(x)$

$$\frac{d}{dx}(\cos u) = -\sin u \frac{du}{dx}$$

or equivalently $\qquad (\cos(f(x)))' = -\sin(f(x))f'(x)$

$$\frac{d}{dx}(\tan u) = \sec^2 u \frac{du}{dx}$$

or equivalently $\qquad (\tan(f(x)))' = \sec^2(f(x))f'(x)$

$$\frac{d}{dx}(\cot u) = -\csc^2 u \frac{du}{dx}$$

or equivalently $\qquad (\cot(f(x)))' = -\csc^2(f(x))f'(x)$

$$\frac{d}{dx}(\sec u) = \sec u \tan u \frac{du}{dx}$$

or equivalently

$$(\sec(f(x)))' = \sec(f(x))\tan(f(x))f'(x)$$

$$\frac{d}{dx}(\csc u) = -\csc u \cot u \frac{du}{dx}$$

or equivalently

$$(\csc(f(x)))' = -\csc(f(x))\cot(f(x))f'(x)$$

Derivatives of exponential functions

$$\frac{d}{dx}(e^x) = e^x$$

$$\frac{d}{dx}(a^x) = a^x \ln a \quad , \quad a = \text{constant}$$

$$\frac{d}{dx}(e^u) = e^u \frac{du}{dx}$$

or equivalently $\quad (e^{f(x)})' = e^{f(x)}f'(x)$

$$\frac{d}{dx}(a^u) = a^u \ln a \frac{du}{dx} \quad , \quad a = \text{constant}$$

or equivalently $\quad (a^{f(x)})' = a^{f(x)}\ln a \, f'(x)$

Derivatives of logarithmic functions

$$\frac{d}{dx}(\ln x) = \frac{1}{x} \quad , \quad x \neq 0$$

$$\frac{d}{dx}(\log_a x) = \frac{\log_a e}{x} \quad , \quad a \neq 0,1$$

$$\frac{d}{dx}(\ln u) = \frac{1}{u}\frac{du}{dx} \quad , \quad u \neq 0$$

or equivalently $\qquad (\ln(f(x)))' = \dfrac{1}{f(x)}f'(x) \quad ,$

$f(x) \neq 0$

$$\frac{d}{dx}(\log_a u) = \frac{\log_a e}{u}\frac{du}{dx} \quad , \quad u \neq 0$$

or equivalently $\qquad (\log_a(f(x)))' = \dfrac{\log_a e}{f(x)}f'(x) \quad ,$

$f(x) \neq 0 \quad , \quad a \neq 0,1$

Derivatives of hyperbolic functions

$$\frac{d}{dx}(\sinh x) = \cosh x$$

$$\frac{d}{dx}(\cosh x) = \sinh x$$

$$\frac{d}{dx}(\tanh x) = \sec h^2 x$$

$$\frac{d}{dx}(\coth x) = -\csc h^2 x$$

$$\frac{d}{dx}(\sec h\,x) = -\sec h\,x \tanh x$$

$$\frac{d}{dx}(\csc h\,x) = -\csc h\,x \coth x$$

$$\frac{d}{dx}(\sinh u) = \cosh u\,\frac{du}{dx}$$

or equivalently $\qquad (\sinh(f(x)))' = \cosh(f(x))f'(x)$

$$\frac{d}{dx}(\cosh u) = \sinh u\,\frac{du}{dx}$$

or equivalently $\qquad (\cosh(f(x)))' = \sinh(f(x))f'(x)$

$$\frac{d}{dx}(\tanh u) = \sec h^2 u\,\frac{du}{dx}$$

or equivalently $\qquad (\tanh(f(x)))' = \sec h^2(f(x))f'(x)$

$$\frac{d}{dx}(\coth u) = -\csc h^2 u\,\frac{du}{dx}$$

or equivalently $\qquad (\coth(f(x)))' = -\csc h^2(f(x))f'(x)$

$$\frac{d}{dx}(\sec h\,u) = -\sec h\,u \tanh u\,\frac{du}{dx}$$

or equivalently

$$\left(\sec h(f(x))\right)' = -\sec h(f(x))\tanh(f(x))f'(x)$$

$$\frac{d}{dx}(\csc hu) = -\csc hu \coth u \frac{du}{dx}$$

or equivalently

$$\left(\csc h(f(x))\right)' = -\csc h(f(x))\coth(f(x))f'(x)$$

Higher derivatives

$$y'' = \frac{d}{dx}\left(\frac{dy}{dx}\right) = \frac{d^2 y}{dx^2} = f''(x) \qquad \text{Second derivative}$$

$$y''' = \frac{d}{dx}\left(\frac{d^2 y}{dx^2}\right) = \frac{d^3 y}{dx^3} = f'''(x) \qquad \text{Third derivative}$$

$$y^{iv} = \frac{d}{dx}\left(\frac{d^3 y}{dx^3}\right) = \frac{d^4 y}{dx^4} = f^{iv}(x) \qquad \text{Fourth derivative}$$

Differentials

$$dy = f'(x)dx$$

Rules for differentials

$$d(u + v) = du + dv$$

$$d(u - v) = du - dv$$

$$d(uv) = u\,dv + v\,du$$

$$d\left(\frac{u}{v}\right) = \frac{v\,du - u\,dv}{v^2} \quad , \qquad v \neq 0$$

$$d(u^n) = nu^{n-1}\,du$$

$$d(\sin u) = \cos u\,du$$

$$d(\cos u) = -\sin u\,du$$

$$d(\tan u) = \sec^2 u\,du$$

Partial derivatives

Definition of partial derivatives

$$\frac{\partial f}{\partial x} = \lim_{\Delta x \to 0} \frac{f(x + \Delta x, y) - f(x, y)}{\Delta x}$$

$$\frac{\partial f}{\partial y} = \lim_{\Delta y \to 0} \frac{f(x, y + \Delta y) - f(x, y)}{\Delta y}$$

Second partial derivatives

$$\frac{\partial^2 f}{\partial x^2} = \frac{\partial}{\partial x}\left(\frac{\partial f}{\partial x}\right)$$

$$\frac{\partial^2 f}{\partial y^2} = \frac{\partial}{\partial y}\left(\frac{\partial f}{\partial y}\right)$$

$$\frac{\partial^2 f}{\partial x \partial y} = \frac{\partial}{\partial x}\left(\frac{\partial f}{\partial y}\right)$$

$$\frac{\partial^2 f}{\partial y \partial x} = \frac{\partial}{\partial y}\left(\frac{\partial f}{\partial x}\right)$$

The differential of a function

$$df = \frac{\partial f}{\partial x}dx + \frac{\partial f}{\partial y}dy$$

13. Indefinite Integrals

In the following formulas, note that the constant of integration is not shown. Furthermore, note that $a = $ constant and $n = $ constant.

$$\int a\,dx = ax$$

$$\int ax\,dx = \frac{1}{2}ax^2$$

$$\int ax^n\,dx = \frac{ax^{n+1}}{n+1} \quad , \quad n \neq -1$$

$$\int \frac{a}{x}\,dx = a\ln|x|$$

$$\int af(x)\,dx = a\int f(x)\,dx$$

$$\int (f(x)+g(x))dx = \int f(x)dx + \int g(x)dx$$

$$\int (f(x)-g(x))dx = \int f(x)dx - \int g(x)dx$$

Integration by Parts using this notation

$$(u = f(x), \quad v = g(x), \quad du = f'(x)dx, \quad dv = g'(x)dx)$$

$$\int u\,dv = uv - \int v\,du$$

Other integration formulas

$$\int f(ax)dx = \frac{1}{a}\int f(w)dw$$

$$\int g(f(x))dx = \int \frac{g(w)}{f'(x)}dw, \quad w = f(x)$$

$$\int w^{n}dw = \frac{w^{n+1}}{n+1}, \quad n \neq -1$$

$$\int \frac{dw}{w} = \ln|w|$$

Indefinite integrals of exponential functions

$$\int e^{w}dw = e^{w}$$

$$\int a^{w}dw = \frac{a^{w}}{\ln a}, \quad a > 0,\ a \neq 1$$

Indefinite integrals of trigonometric functions

$$\int \sin w \, dw = -\cos w$$

$$\int \cos w \, dw = \sin w$$

$$\int \tan w \, dw = -\ln(\cos w)$$

$$\int \cot w \, dw = \ln(\sin w)$$

$$\int \sec w \, dw = \ln(\sec w + \tan w)$$

$$\int \csc w \, dw = \ln(\csc w - \cot w)$$

$$\int \sec^2 w \, dw = \tan w$$

$$\int \csc^2 w \, dw = -\cot w$$

$$\int \tan^2 w \, dw = \tan w - w$$

$$\int \cot^2 w \, dw = -\cot w - w$$

$$\int \sin^2 w \, dw = \frac{1}{2}(w - \sin w \cos w)$$

$$\int \cos^2 w \, dw = \frac{1}{2}(w + \sin w \cos w)$$

$$\int \sec w \tan w \, dw = \sec w$$

$$\int \csc w \cot w \, dw = -\csc w$$

$$\int \sin(ax)\,dx = -\frac{1}{a}\cos(ax)$$

$$\int \cos(ax)\,dx = \frac{1}{a}\sin(ax)$$

Indefinite integrals of hyperbolic functions

$$\int \sinh w \, dw = \cosh w$$

$$\int \cosh w \, dw = \sinh w$$

$$\int \tanh w \, dw = \ln(\cosh w)$$

$$\int \coth w \, dw = \ln(\sinh w)$$

$$\int \sec h\, w \, dw = \sin^{-1}(\tanh w) = 2\tan^{-1}(e^{w})$$

$$\int \csc h\, w \, dw = \ln\left(\tanh\frac{w}{2}\right) = -\coth^{-1}(e^{w})$$

$$\int \sec h^{2} w \, dw = \tanh w$$

$$\int \csc h^{2} w \, dw = -\coth w$$

$$\int \tanh^2 w\, dw = w - \tanh w$$

$$\int \coth^2 w\, dw = w - \coth w$$

$$\int \sinh^2 w\, dw = \frac{1}{2}(\sinh w \cosh w - w)$$

$$\int \cosh^2 w\, dw = \frac{1}{2}(\sinh w \cosh w + w)$$

$$\int \sec h\, w \tanh w\, dw = -\sec h\, w$$

$$\int \csc h\, w \coth w\, dw = -\csc h\, w$$

Integration by substitution

$$\int F(ax+b)\,dx = \frac{1}{a}\int F(w)\,dw \qquad , \qquad w = ax + b$$

$$\int F(\sqrt{ax+b})\,dx = \frac{2}{a}\int wF(w)\,dw \quad , \qquad w = \sqrt{ax+b}$$

$$\int F(\sqrt{a^2-x^2})\,dx = a\int F(a\cos w)\cos w\, dw \quad , \qquad w = a\sin x$$

$$\int F(\sqrt{x^2+a^2})\,dx = a\int F(a\sec w)\sec^2 w\, dw \quad , \qquad w = a\tan x$$

$$\int F(\sqrt{x^2-a^2})\,dx = a\int F(a\tan w)\sec w \tan w\, dw , \quad w = a\sec x$$

$$\int F(e^{ax})\,dx = \frac{1}{a}\int \frac{F(w)}{w}\,dw \qquad , \qquad w = e^{ax}$$

$$\int F(\ln x)\,dx = \int F(w)e^{w}\,dw \quad , \qquad w = \ln x$$

14. Definite Integrals

Definition of the definite integral

$$\int_{a}^{b} f(x)\,dx = \lim_{n\to\infty}[f(a)\Delta x + f(a+\Delta x)\Delta x + f(a+2\Delta x)\Delta x$$
$$+ \ldots\ldots + f(a+(n-1)\Delta x)\Delta x]$$

where the interval $[a,b]$ is subdivided into n equal parts of length $\Delta x = \dfrac{b-a}{n}$.

Properties of the definite integral

$$\int_{a}^{b} f(x)\,dx = g(x)\Big]_{a}^{b} = g(b) - g(a)$$

where $f(x) = \dfrac{d}{dx}g(x) = g'(x)$.

$$\int_{a}^{b}[f(x)+g(x)]\,dx = \int_{a}^{b}f(x)\,dx + \int_{a}^{b}g(x)\,dx$$

$$\int_{a}^{b}[f(x)-g(x)]\,dx = \int_{a}^{b}f(x)\,dx - \int_{a}^{b}g(x)\,dx$$

$$\int_{a}^{b}cf(x)\,dx = c\int_{a}^{b}f(x)\,dx$$

$$\int_a^a f(x)dx = 0$$

$$\int_a^b f(x)dx = -\int_b^a f(x)dx$$

$$\int_a^b f(x)dx = \int_a^c f(x)dx + \int_c^b f(x)dx$$

Mean Value Theorem

$$\int_a^b f(x)dx = (b-a)f(c) \quad , \quad \text{where } c \text{ is between } a \text{ and } b$$
and $f(x)$ is continuous in $[a,b]$.

Approximation of definite integrals

Subdivide the interval $[a,b]$ into n equal parts by the points
$a = x_0, x_1, x_2, x_3, \ldots\ldots, x_{n-1}, x_n = b$ where

$$y_i = f(x_i), \quad i = 0,1,2,3,\ldots,(n-1),n \quad \text{and } h = \frac{b-a}{n}.$$

Rectangular rule of approximation

$$\int_a^b f(x)dx \approx h(y_0 + y_1 + y_2 + y_3 + \ldots\ldots + y_{n-1})$$

Trapezoidal rule of approximation

$$\int_a^b f(x)dx \approx \frac{h}{2}(y_0 + 2y_1 + 2y_2 + 2y_3 + \ldots\ldots + 2y_{n-1} + y_n)$$

Simpsons' rule of approximation

$$\int_a^b f(x)dx \approx \frac{h}{3}\left(y_0 + 4y_1 + 2y_2 + 4y_3 + \ldots\ldots + 2y_{n-2} + 4y_{n-1} + y_n\right)$$

Selected definite integrals

$$\int_0^\pi \sin(mx)\sin(nx)dx = \begin{cases} 0 & , \ m \neq n \\ \dfrac{\pi}{2} & , \ m = n \end{cases}$$

$$\int_0^\pi \cos(mx)\cos(nx)dx = \begin{cases} 0 & , \ m \neq n \\ \dfrac{\pi}{2} & , \ m = n \end{cases}$$

$$\int_0^\pi \sin(mx)\cos(nx)dx = \begin{cases} 0 & , \ m+n \ \ even \\ \dfrac{2m}{m^2 - n^2} & , \ m+n \ \ odd \end{cases}$$

$$\int_0^{\frac{\pi}{2}} \sin^2 x\, dx = \int_0^{\frac{\pi}{2}} \cos^2 x\, dx = \frac{\pi}{4}$$

$$\int_0^\infty \frac{\sin x}{x}dx = \frac{\pi}{2}$$

$$\int_0^\infty \frac{\sin x \cos x}{x} = \frac{\pi}{4}$$

$$\int_0^\infty \frac{\tan x}{x} = \frac{\pi}{2}$$

15. Series

Arithmetic series

$$a + (a+d) + (a+2d) + ... + (a + (n-1)d) = \frac{n(a+l)}{2}$$

where $l = a + (n-1)d$.

Examples of arithmetic series

$$1 + 2 + 3 + ... + n = \frac{n(n+1)}{2}$$

$$1 + 3 + 5 + ... + (2n-1) = n^2$$

Geometric series

$$a + ar + ar^2 + ar^3 + ... + ar^{n-1} = a\left(\frac{1-r^n}{1-r}\right), \; r \neq 1$$

Example of a geometric series

$$a + ar + ar^2 + ar^3 + ... + ar^{n-1} = \frac{a}{1-r} \; , \quad -1 < r < 1$$

Examples of other types of series

$$1^2 + 2^2 + 3^2 + ... + n^2 = \frac{n(n+1)(2n+1)}{6}$$

$$1^3 + 2^3 + 3^3 + \ldots\ldots\ldots + n^3 = \frac{n^2(n+1)^2}{4} = (1 + 2 + 3 + \ldots + n)^2$$

$$1 - \frac{1}{2} + \frac{1}{3} - \frac{1}{4} + \frac{1}{5} - \ldots\ldots\ldots = \ln 2$$

$$1 - \frac{1}{3} + \frac{1}{5} - \frac{1}{7} + \frac{1}{9} - \ldots\ldots\ldots = \frac{\pi}{4}$$

$$\frac{1}{1^2} + \frac{1}{2^2} + \frac{1}{3^2} + \frac{1}{4^2} + \ldots\ldots\ldots + \frac{1}{n^2} = \frac{\pi^2}{6}$$

$$\frac{1}{1^2} - \frac{1}{2^2} + \frac{1}{3^2} - \frac{1}{4^2} + \ldots\ldots\ldots = \frac{\pi^2}{12}$$

$$\frac{1}{1^2} + \frac{1}{3^2} + \frac{1}{5^2} + \frac{1}{7^2} + \ldots\ldots\ldots = \frac{\pi^2}{8}$$

$$\frac{1}{1\cdot 3} + \frac{1}{3\cdot 5} + \frac{1}{5\cdot 7} + \frac{1}{7\cdot 9} + \ldots\ldots\ldots = \frac{1}{2}$$

$$\frac{1}{1\cdot 3} + \frac{1}{2\cdot 4} + \frac{1}{3\cdot 5} + \frac{1}{4\cdot 6} + \ldots\ldots\ldots = \frac{3}{4}$$

16. Taylor Series

$$f(x) = f(a) + f'(a)(x-a) + \frac{1}{2!}f''(a)(x-a)^2 + \ldots\ldots$$

$$+ \frac{1}{(n-1)!}f^{(n-1)}(a)(x-a)^{n-1} + R_n$$

where the remainder R_n is given by the following equation

$$R_n = \frac{1}{n!}f^{(n)}(\xi)(x-a)^n$$

and ξ is between a and x. Also, the function f must be continuous with continuous derivatives.

The infinite series given above is called the *Taylor series* for $f(x)$ about $x = a$ if $\lim_{n\to\infty} R_n = 0$.

If $a = 0$, then the series is called a *Maclaurin series.*

Examples of Taylor and Maclaurin series

$$\frac{1}{1+x} = 1 - x + x^2 - x^3 + x^4 - \ldots\ldots\ldots\ , \quad -1 < x < 1$$

$$\sqrt{1+x} = 1 + \frac{1}{2}x - \frac{1}{2\cdot4}x^2 + \frac{1\cdot3}{2\cdot4\cdot6}x^3 - \ldots\ldots\ , \quad -1 < x \le 1$$

$$e^x = 1 + x + \frac{x^2}{2!} + \frac{x^3}{3!} + \ldots\ldots\ldots\ , \quad -\infty < x < \infty$$

$$a^x = 1 + x\ln a + \frac{(x\ln a)^2}{2!} + \frac{(x\ln a)^3}{3!} + \ldots\ldots\ldots\ , \quad -\infty < x < \infty$$

$$\ln(1+x) = x - \frac{x^2}{2} + \frac{x^3}{3} - \frac{x^4}{4} + \ldots\ldots\ , \quad -1 < x \le 1$$

$$\sin x = x - \frac{x^3}{3!} + \frac{x^5}{5!} - \frac{x^7}{7!} + \ldots\ldots\ , \quad -\infty < x < \infty$$

$$\cos x = 1 - \frac{x^2}{2!} + \frac{x^4}{4!} - \frac{x^6}{6!} + \ldots\ldots\ , \quad -\infty < x < \infty$$

$$\sinh x = x + \frac{x^3}{3!} + \frac{x^5}{5!} + \frac{x^7}{7!} + \ldots\ldots\ , \quad -\infty < x < \infty$$

$$\cosh x = 1 + \frac{x^2}{2!} + \frac{x^4}{4!} + \frac{x^6}{6!} + \ldots\ldots\ , \quad -\infty < x < \infty$$

17. Inequalities

Triangle inequality

$$\left|a_1\right| - \left|a_2\right| \le \left|a_1 + a_2\right| \le \left|a_1\right| + \left|a_2\right|$$

$$\left|a_1 + a_2 + a_3 + \ldots\ldots + a_n\right| \le \left|a_1\right| + \left|a_2\right| + \left|a_3\right| + \ldots\ldots + \left|a_n\right|$$

18. Conversion Factors

Length

1 kilometer = 1000 meters

1 meter = 100 centimeters

1 meter – 1000 millimeters

1 centimeter = 0.01 meter

1 millimeter = 0.001 meter

1 inch = 2.540 centimeters

1 foot = 30.48 centimeters

1 mile = 1.609 kilometers

1 centimeter = 0.3937 inch

1 meter = 39.37 inches

1 kilometer = 0.6214 mile

Volume

1 liter = 1000 cubic centimeters

1 cubic meter = 1000 liters

1 cubic foot = 28.32 liters

Mass

1 kilogram = 1000 grams

1 kilogram = 2.2046 pounds

1 pound = 453.6 grams

For detailed conversions of other units, visit the website www.onlineconversion.com for immediate online conversions.

19. List of Prime Numbers

The following table lists the first 30 prime numbers:

2	3	5	7
11	13	17	19
23	29	31	37
41	43	47	53
59	61	67	71
73	79	83	89
97	101	103	107
109	113		

References

1. Spiegel, M. R., *Schaum's Mathematical Handbook of Formulas and Tables*, McGraw-Hill, Second Edition, 1998.

2. Spiegel, M. R., Lipschutz, S. and Liu, J., *Schaum's Outline of Mathematical Handbook of Formulas and Tables* , McGraw-Hill, Third Edition, 2008.

3. Jeffrey, A. and Dai, H. H., *Handbook of Mathematical Formulas and Integrals*, Academic Press, Fourth Edition, 2008.

4. Abramowitz, M. and Stegun, I. A. , *Handbook of Mathematical Functions: With Formulas, Graphs, and Mathematical Tables* , Dover Publications, 1965.

5. Zwillinger, D. , *CRC Standard Mathematical Tables and Formulas*, Chapman & Hall/CRC, 31st Edition, 2002.

6. Tallarida, R. J. , *Pocket Book of Integrals and Mathematical Formulas*, Chapman & Hall/CRC, 2008.

7. Ballast, D. , *Architect's Handbook of Formulas, Tables, and Mathematical Calculations*, Prentice Hall, 1988.

8. Burington, R. S., *Handbook of Mathematical Tables and Formulas*, McGraw-Hill, Fifth Edition, 1973.

9. Luderer, B., Nollau, V. and Vetters, K. , *Mathematical Formulas for Economists*, Springer, Third Edition, 2006.

10. Spiegel M. R. and Liu, J. , *Schaum's Easy Outline of Mathematical Handbook of Formulas and Tables,* , McGraw-Hill, 2001.

11. Boljanovic, V. , *Applied Mathematical and Physical Formulas Pocket Reference*, Industrial Press, 2006.

12. Bartsch, H. J. , *Handbook of Mathematical Formulas*, Academic Press, 1974.

Internet Links

1. S. O. S. Mathematics: Tables and Formulas

   http://www.sosmath.com/tables/tables.html

2. The World of Math Online – Math.com
   Math Reference Tables

   http://www.math.com/tables/

3. Measurement Formulas: Geometry

   http://library.thinkquest.org/2647/geometry/measur e/measure.htm

4. Math Formulas, Math Tables

   http://math.about.com/od/formulas/Math_Formula s_Math_Tables.htm

5. The Educational Encyclopedia: Mathematical Formulas

   http://www.educypedia.be/education/mathematicsfo rmulas.htm

6. Mathematical Formulas for Areas and Volumes

   http://artstuf.com/mathmatical-formulas.html

7. Commonly Used Mathematical Formulas

   http://campus.northpark.edu/math/PreCalculus/formulas.html

Notes

Notes